START WITH A QUESTION

What Is a Goldilocks Planet?

FAQs About Space

Rachel Stuckey

A Crabtree Forest Book

Crabtree Publishing
crabtreebooks.com

Notes for Readers and Class Discussion

This book is designed to teach readers about core subject areas, build curiosity, and inspire further investigation. Readers are encouraged to build upon what they already know about the subject and engage in topics they want to learn more about. Here are a few guiding questions to prompt reflection and discussion. Possible answers appear in red.

Before Reading:

Read the title and look at the table of contents. What do you already know about space?

- *I know that there are planets, stars, and other objects in space.*
- *I know that the planets in our solar system orbit the Sun.*

What would you like to learn about space?

- *I would like to learn about what lies beyond our solar system.*
- *I would like to learn how other planets compare to Earth.*

During Reading:

Pause after reading each page or chapter. What questions do you have about what you read? What are you curious to know?

- *I wonder if it will ever be possible to learn if the universe has an end.*
- *I am curious to know if there is life on other planets.*

Make connections to what you are reading. How is this like something you already know?

- *I know that astronauts travel to space to learn more about it.*
- *I know that there are many stars in the night sky and that the Sun is a star.*

After Reading:

Recall key details about the book. What was the author trying to teach readers?

- *The author was trying to teach readers that there are still many unanswered questions about space.*
- *The author was trying to teach readers that scientists have learned a lot about our solar system and the universe.*

How did the images and captions help you understand more?

- *The photographs helped me see different space objects.*
- *The captions helped me understand more details about space.*

Table of Contents

Think Like a Scientist

People have been asking questions about space throughout history. Ancient peoples observed the stars and planets and told stories about the things they saw. In 1609, the Italian artist and inventor Galileo Galilei designed a powerful **telescope** and used it to look at space. He saw mountains and craters on the Moon. He saw the surface of the Sun and planets with moons and rings.

Since then, scientists asked more questions about space. They conducted investigations and built bigger and better telescopes to find answers to their questions. Their investigations often led them in new directions. Scientists learned that the eight planets in our solar system **orbit** the Sun. They discovered stars and **galaxies** that were too far away to see in the night sky. They also learned about other space objects, such as **asteroids** and **comets**.

Over decades of space investigations, people have sent thousands of **satellites** into the sky, humans to the Moon, a telescope to orbit the Sun, and **probes** to the edge of our solar system. Space science is always evolving as we learn more about the universe. Perhaps the questions and answers in this book will inspire you to investigate more questions about space—just like a scientist.

The planets of the inner solar system, such as Earth, are made of rock. The planets farther out, such as Uranus, are made of mostly gas or ice.

What questions about space spark your interest?

Where does the sky end and space begin?

Earth is surrounded by layers of gases known as the **atmosphere**. The atmosphere does not end abruptly where space begins, however. Instead, it becomes thinner and thinner the higher it gets from Earth's surface. This is because Earth's **gravity** is no longer strong enough to hold onto the gas **molecules** in its atmosphere.

Many scientists say that space begins at the Kármán line, which is about 62 miles (100 km) above Earth's surface. Though Earth's atmosphere extends beyond this point, the air is extremely thin. Regular airplanes cannot fly in the upper atmosphere because there is not enough air pressure to create lift—or enough oxygen for people to breathe. Not all experts agree, however, that the Kármán line is where space begins. For example, some say that space begins where an object would begin to orbit Earth, as low as 50 miles (80 km) above Earth's surface.

Gravity keeps us from floating into space because it pulls us toward Earth. The force of gravity is so strong, in fact, that we need rockets to travel out of the atmosphere and into space.

Above the Kármán line, the sky begins to turn from blue to black. That's because there are too few gas molecules to scatter the Sun's light, which creates the blue color we see.

Where does space end?

It's hard to imagine how big space really is. Our own solar system is about 186 billion miles (300 billion km) from one side to the other. Just like the Kármán line, the edge of our solar system is where the Sun's gravity—which keeps planets and space objects in orbit around it—loses strength. The area between the edge of our solar system and other star systems in our galaxy, or the Milky Way, is called interstellar space.

Scientists estimate that there are between 100 and 400 billion stars in the Milky Way. But space is much more than the Milky Way. Scientists estimate that there are between 2 and 10 trillion galaxies in the universe. We can't see all these galaxies, so scientists use advanced mathematics to make these estimates. Scientists don't know where the universe ends—or if it ends at all. But they can estimate the **diameter** of the observable universe, or the universe they can see: 93 billion light-years!

DID YOU KNOW?

Light is fast—in one minute, light can travel more than 11 million miles (17.7 million km). But a light-year is a measure of distance, not time. It is how people measure the immense distances in space. A light-year is the distance light can travel in one year on Earth, or 365 days. One light-year equals 5.88 trillion miles (9.46 trillion km). You would have to travel around the world more than 200 million times to travel the distance of one light-year!

Andromeda is a galaxy about 2.5 million light-years away from Earth. To the human eye, it looks like a star in the night sky. With a telescope, it is possible to see its elliptical, or oval, shape.

Earth is about eight light-minutes from the Sun.

If it's not empty, why do we call it "space"?

For centuries, we have used the word "space" or "outer space" to describe what's beyond Earth. Space may have seemed empty because we did not always know what it contained. But now we know there is a difference between space and the universe. The universe includes everything—the stars, planets, space objects, clouds of gas and dust—and all of space. Today, scientists and astronauts use the word "space" to describe the emptiness between all these things.

DID YOU KNOW?

There is no sound in space! Sound is a type of energy that vibrates matter**. When you clap your hands, you vibrate the air molecules around you and the energy travels through the air to vibrate in our ears. But there are no air molecules in space. That means without air molecules to transmit, or pass on, sound energy, there is no sound.**

Space appears dark because there is very little matter for light to bounce off of and reach our eyes.

A nebula is a giant cloud of dust and gas in space. The Orion Nebula is one of the brightest in the Milky Way. It can be seen in the night sky!

Where do planets come from?

Scientists suggest that planets probably begin as tiny grains of dust, smaller than the width of a human hair! There are many steps in the planet formation process. First, dust and gases are pulled together by gravity. They form a cloud called a nebula. As the nebula gets bigger and heavier, it spins and collapses into a clump. Gravity makes it spin faster and faster, heating up as it spins. The heat energy in the center creates a star. Then the remaining dust and gas form a spinning disk around the young star. The matter crashes together and gravity pulls it into spinning clumps. Eventually, these spinning clumps pull in more matter and get bigger and bigger. They become planets that orbit around the star.

Why is a year on Mars longer than a year on Earth ?

Just like Earth's gravity pulls us toward the ground, planets in the solar system are being pulled toward the Sun. As the object with the most **mass** in the solar system, it has the strongest force of gravity. This keeps the planets from drifting off into space. The constant movement of planets as they orbit the Sun is what keeps them from being pulled into it. For any planet, the number of days it takes to orbit the Sun equals one year. Earth takes 365 days to orbit the Sun.

Mars is smaller than Earth, at one-tenth its mass. That means Mars has less gravity than Earth. But because Mars is farther away from the Sun, its orbit takes longer. A year on Mars is 687 Earth days. A planet's day is measured by the amount of time it takes to make a full rotation on its **axis**. On Earth, a day is about 24 hours. A day on Mars is slightly longer: 24.65 hours. Scientists call a solar day on Mars a "sol" to avoid confusing it with a "day" on Earth.

Earth is traveling about 67,000 miles per hour (107,826 km/h) in its orbit around the Sun. Mars is traveling around 53,979 miles per hour (86,871 km/h).

Mars and Earth are part of the inner solar system—the name for the four planets closest to the Sun. These four planets are **terrestrial** planets. Mercury and Venus are the other two planets in the inner solar system. But they have key differences from Earth.

Mercury is the smallest planet, at a little over 3,000 miles (4,800 km) across. That's less than the distance from New York to Hawaii! Mercury has no atmosphere to protect it from the Sun's energy or trap its heat. That means the temperature on Mercury can reach 800 degrees Fahrenheit (427 degrees Celsius) during the day and –290 degrees Fahrenheit (–179 degrees Celsius) at night.

Venus is similar in size to Earth, but because it is a little closer to the Sun, it only takes 225 days to orbit. But Venus also rotates very slowly. A day on Venus is 243 Earth days long. That means a day is longer than a year on Venus!

The inner solar system also has millions of smaller rocks, called asteroids, left over from planet formation. Gravity pulls most of these rocks toward each other into the asteroid belt, the area between Mars and Jupiter where most asteroids are found.

Mars is called the "red planet" because it has rusty dust in its atmosphere, which makes it appear red in color.

When the solar system formed, the lighter materials like gases were pushed farther away and the heavy materials like rock stayed closer to the Sun. That makes the inner four planets terrestrial, or rocky.

Why does the Moon seem to change size and shape?

The Moon completes a full orbit of Earth every 27 days. It has an elliptical orbit. That means that at certain times, it is closer to Earth. But the Moon is too far away for us to see the difference in its distance from Earth. So when the Moon sometimes looks bigger, it's because our minds are playing a trick on us called the "Moon illusion." Our brains compare the size of the Moon to the size of other objects we can see. If we are near a mountain, for example, the Moon looks small in comparison. But if that mountain was far away, it would look smaller, and the Moon would appear larger.

The shape of the part of the Moon we can see is called a Moon phase. Together, all the shapes in a month are called the phases of the Moon.

The Moon never really changes its shape. But shadows make it look like it has. The Moon is visible to us because light from the Sun shines on it. As the Moon orbits Earth, we see different areas of its surface illuminated by the Sun. During a full Moon, for example, the Sun is shining directly on the side that faces Earth.

DID YOU KNOW?

Ocean tides are caused by the Moon's gravity. Though the Moon is small in comparison to Earth, its gravity is strong enough to pull Earth's oceans into a bulge. As Earth rotates on its axis, the position of the bulge changes throughout each day. When a sea or ocean is bulging, it is in high tide**. At low tide, a sea or ocean is not bulging and the water is farther from the shoreline.**

The oceans bulge on the sides closest to and farthest from the Moon at any point. That's because of the way the Moon's gravity affects all of Earth—the force puts pressure on the water until it bulges out the sides.

How can planets be made of gas?

Jupiter and Saturn are known as gas giants. They are made up mostly of hydrogen and helium gases, though they probably have solid cores, or centers. Because these planets are farther from the Sun, their gravity draws in the lighter materials in the solar system, like these gases. The gases are packed so tightly by the gravity of the planet's core that they form a kind of ocean that makes up the planet's surface. Then a lighter layer of gases forms its atmosphere. Neptune and Uranus are also giants, but instead of hydrogen and helium they are formed mostly from water, **ammonia**, and **methane**. Though these planets are sometimes called ice giants, they are mostly made up of liquids.

These four planets of the outer solar system are called giants for a reason. Uranus, Neptune, and Saturn are 14, 17, and 95 times the size of Earth, respectively. But Jupiter is the true giant—it is 300 times the size of Earth.

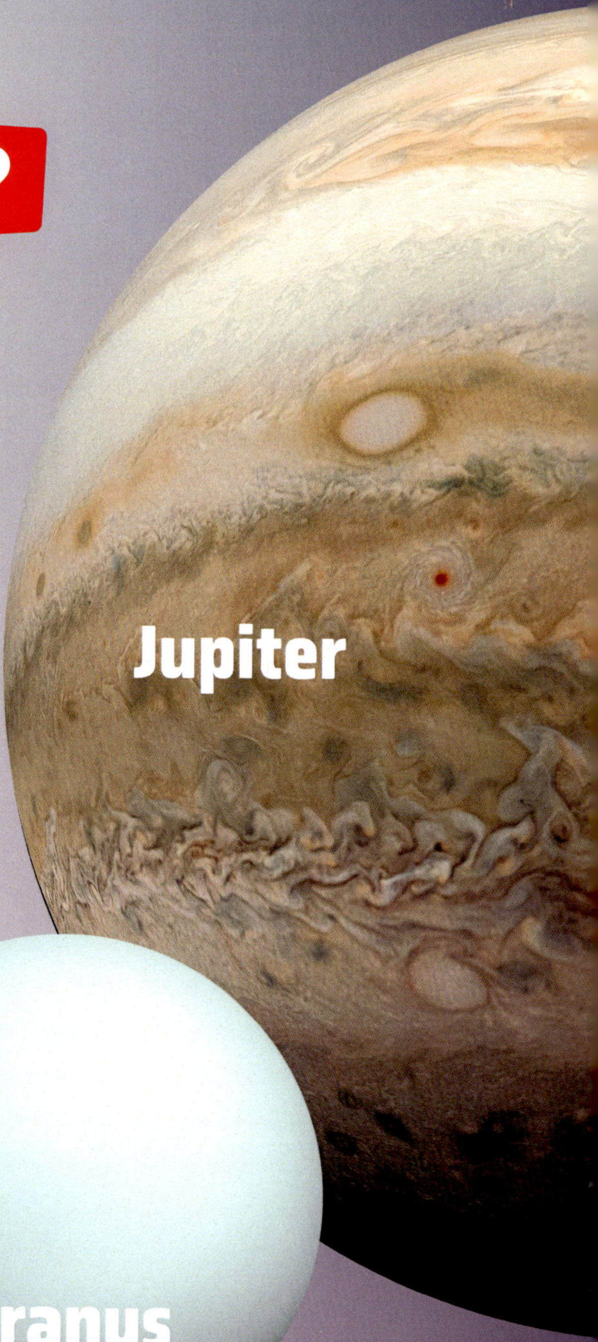

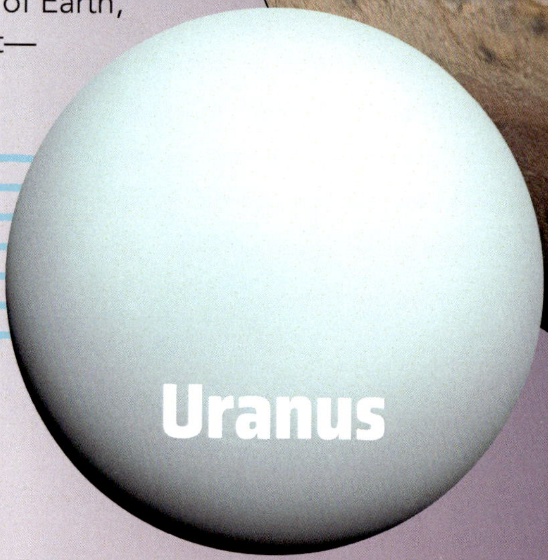

Neptune, the eighth planet in our solar system, is about 2.8 billion miles (4.5 billion km) away from the Sun. It takes light four hours to travel from the Sun to Neptune.

DID YOU KNOW?

Saturn has the most moons of any planet in our solar system. In 1610, Galileo discovered four moons around Jupiter using a simple telescope. In 1655, an astronomer named Christiaan Huygens discovered Titan, Saturn's largest moon. As telescope technology improved, we saw more and more moons orbiting both planets. As of 2025, astronomers have confirmed that Saturn's total is 274 and Jupiter's total is 95.

The debris particles in Saturn's rings range in size from tiny grains of sand to large chunks the size of mountains!

Where did Saturn's rings come from?

Saturn is known for its rings, which Galileo first saw in 1610 through his telescope. But we now know that all the giant planets have ring systems. Saturn's rings are made mostly of ice, so they reflect light from the Sun. This makes them easier to see through a simple telescope. Jupiter's rings are harder to see because they are made of dust. Uranus and Neptune's rings are dark in color. The rings around these planets are so faint that they are difficult to see, even with a telescope.

Planetary rings are created when a planet's gravity pulls a cloud of **debris** toward it and into orbit. The debris in Saturn's rings may have come from an asteroid crashing into a moon, or from comets and asteroids that were crushed by Saturn's gravity. As the debris in the cloud crashed together, each piece of matter was pulled toward each other until eventually, all the debris flattened into rings.

Saturn has seven main rings. Each is thousands of miles wide, but only up to 330 feet (100 m) thick. They are so thin that they appear to be flat when you view them through a telescope.

Why isn't Pluto a planet?

In 2006, scientists decided that to be a planet, a space object in our solar system must: orbit the Sun, be mostly round, and be large enough that its gravity clears away other large objects in its orbit. They used this criteria to decide that Pluto should be reclassified as a dwarf planet. A dwarf planet is mostly round and orbits the Sun, but due to its size, does not clear its orbit.

In fact, Pluto is smaller than many of the moons in the solar system. These include Neptune's Triton and Jupiter's Callisto, Io, and Europa. Pluto isn't just small—its orbit is also different from the other planets. All the other planets orbit on the same **plane**, or level. Pluto's orbit, on the other hand, is on a different angle around the Sun. It overlaps with similarly sized objects in the Kuiper Belt, which means its gravity is not strong enough to clear its path.

Pluto is not alone. Astronomers have recognized four other dwarf planets in our solar system. Makemake, Haumea, and Eris are dwarf planets found beyond Neptune, like Pluto. Ceres is a dwarf planet in the asteroid belt between Mars and Jupiter.

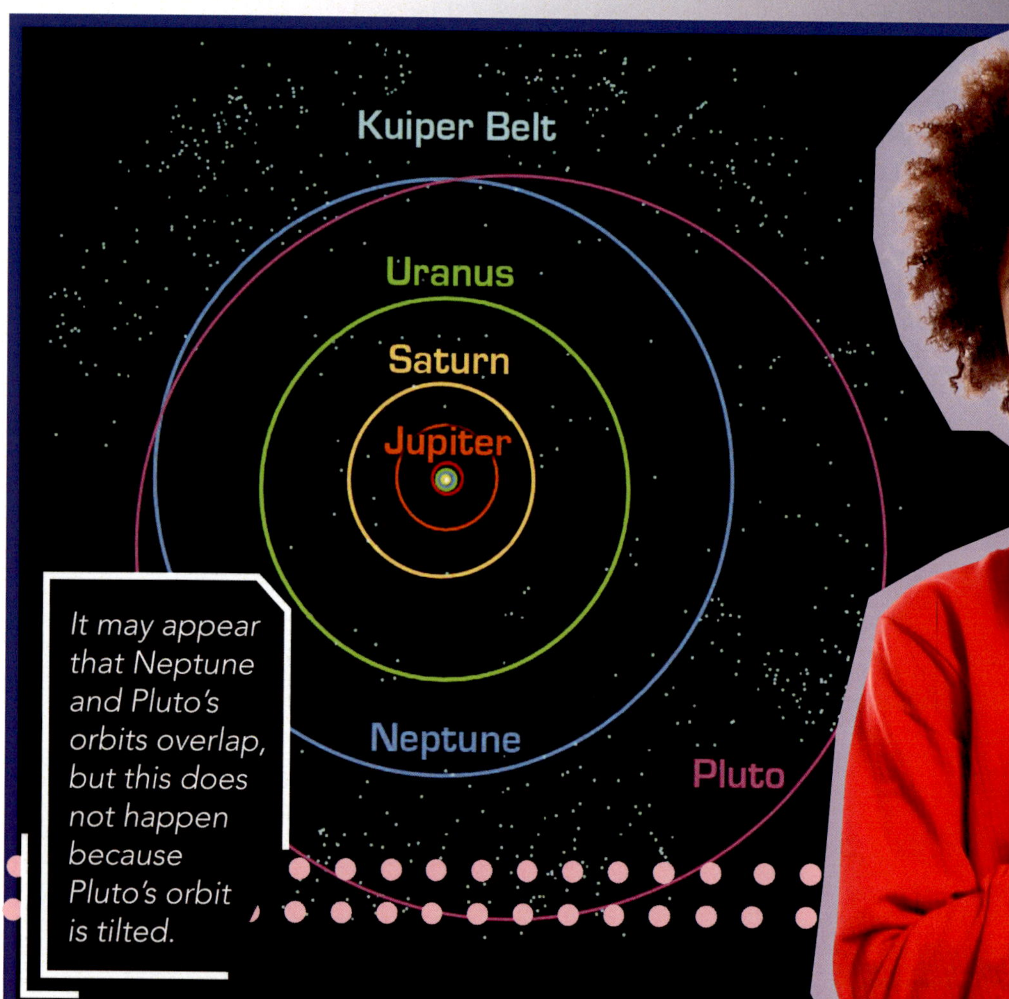

It may appear that Neptune and Pluto's orbits overlap, but this does not happen because Pluto's orbit is tilted.

Pluto's largest moon is Charon. It is so big that the two almost orbit each other as they orbit the Sun.

DID YOU KNOW?

Pluto, Makemake, Eris, and Haumea are part of the Kuiper Belt. This is a doughnut-shaped ring of icy objects that orbits the Sun beyond Neptune. These objects are made of icy ammonia, methane, and water. The strength of Neptune's gravity probably kept all this matter from joining together into another planet.

Why are some stars brighter than others?

The brightness of a star depends on its size and distance. Our Sun is so close—just 8.3 light-minutes away—that it's extremely bright in the sky. The brightest star in the **Northern Hemisphere** is Sirius, which is 8.6 light-years away from us. Sirius is also twice the size of our Sun. Imagine how bright it might look if it were closer to us!

Just like planets, there are giant stars that look brighter and dwarf stars that are more dim. Stars also come in different colors. Betelgeuse, around 650 light-years from Earth, has a warm orange glow. Other stars are white, such as Vega, around 25 light-years away. The color of a star depends on the temperature of the energy it creates. The hottest stars look blue, and the cooler stars appear reddish. Our yellow Sun is in the middle for size and temperature.

DID YOU KNOW?

We can see our own galaxy! Most of the stars we can see are within 10,000 light-years from our Sun. The rest of the millions of stars in our galaxy are so far away, they smush together into a cloudy band. We call this band the Milky Way. Ancient people could see the Milky Way without a telescope. The word "galaxy," in fact, comes from the Greek word for "milk!"

Milky Way

Some of the brightest "stars" you see in the night sky may be planets in our solar system, or faraway galaxies!

Where do stars come from?

We already know that stars are formed from clouds of gas and dust called nebulas. Gravity causes some of the matter in a nebula to clump together and spin, growing bigger and eventually collapsing under the force of its own gravity. The force of this collapse creates a very hot core. The core pulls in and heats up hydrogen and helium gases, slowly creating a star.

Star formation takes millions of years. Once a star is formed, its gravity forms planets from the matter surrounding it. Stars live for billions of years, until they have burned off all their gases. After a star dies, its matter spreads out into the universe, creating new nebulas.

Our star—the Sun—is 4.5 billion years old. Scientists estimate the age of the Sun by testing the age of other things in our solar system, such as Earth. The Sun was formed first, then the planets. Our Sun will probably live for another 5 billion years. As it starts to burn out, it will become a red giant. This means it will get bigger and brighter, but it will also become cooler.

A red giant is a star that has run out of hydrogen and has begun to die. It gets bigger and appears orange.

A star in the regular part of its lifespan is called a main sequence star. Our Sun is a main sequence star. There are different types of these stars.

A red dwarf is the smallest and coolest type of main sequence star. Red dwarves appear orange in color.

A neutron star is the remains of a massive star, much bigger than the Sun. They are extremely **dense**.

A white dwarf is the remains of a star that has burned all of its fuel and shed its atmosphere.

What happens when a star dies?

When a star dies, it explodes. The death of a star is called a supernova. When a star becomes a supernova, the explosion shines brighter than other stars in the sky, sometimes for months. Sometimes supernovas are so bright, we can see them without a telescope. In 2023, a supernova in a nearby galaxy appeared in the sky close to the Big Dipper constellation. It was visible for months through backyard telescopes! Scientists use powerful telescopes to watch hundreds of supernovas in other galaxies every year. The last visible supernova in our galaxy was seen more than 400 years ago.

DID YOU KNOW?

Supernovas are time travelers! When we see stars in the sky, we are looking at light that has traveled for many years before it reaches us on Earth. When a supernova happens in a galaxy that is 20 million light-years away, the light will take 20 million years to reach us. So, when we see a supernova through a telescope, we are actually looking back in time at a 20-million-year-old explosion.

The remnants of supernovas eventually become nebulas, such as the Crab Nebula in the Milky Way (above), where new stars are formed.

What is a constellation?

When human beings first looked at groups of stars in the night sky, before they had telescopes, they drew imaginary lines between them and saw shapes. These shapes are called constellations. Cultures all over the world named the constellations they saw after people, animals, and objects. There are 88 recognized constellations today.

The Big Dipper and the Little Dipper are ***circumpolar*** *constellations that can only be seen in the Northern Hemisphere. These groups of stars look like ladles, or large spoons used to serve liquids.*

Orion the Hunter is a seasonal constellation which can be seen from anywhere on Earth between November and February. You can find Orion in the sky by looking for the three bright stars that make up the hunter's "belt."

The Southern Cross is another circumpolar constellation. It was used by sailors in the ***Southern Hemisphere*** *to help them* ***navigate****.*

What's the difference between an asteroid, a comet, and a meteor

Asteroids and comets are space objects that orbit the Sun. Asteroids are rocks, just like planets of the inner solar system. Comets are balls of rock, dust, and ice. A meteor, on the other hand, is a bright streak of light caused by a meteoroid burning up as it enters a planet's atmosphere. Meteoroids are bits of rock that break off of asteroids, comets, and even the Moon. You might know meteors better by their other name: "shooting stars!"

Most asteroids are found in the asteroid belt. The smallest are less than 33 feet (10 m) wide, while the largest are about 329 miles (529 km) across! Some are large enough to have their own rings and smaller asteroids in orbit around them. These asteroids are too small to be dwarf planets, however.

Scientists have identified more than 1.3 million asteroids. The largest were discovered in the 1800s and are named after the Greek and Roman gods Ceres, Vesta, Pallas, and Hygiea. Ceres is the largest object in the asteroid belt, making up 25 percent of its total mass. Partly because of this, it was reclassified as a dwarf planet in 2006.

Comets are a few miles to tens of miles wide. However, their tails can be millions of miles long!

Sometimes many meteors fall at the same time, creating a meteor shower.

Left over from planet formation, asteroids are some of the oldest space objects in our solar system.

After meteors have fallen to the ground, we call the rocks "meteorites."

Comets travel in elliptical orbits around the Sun. They spend most of their time as icy balls far from the Sun, in the Kuiper Belt and even farther away in the Oort Cloud, beyond the edge of the solar system. When comets get closer to the Sun, the ice starts to melt and creates a glowing halo and tail of dust and gas. Scientists estimate there may be around 1 trillion comets in the outer solar system.

The largest known comet is about 80 miles (129 km) wide. It's about 50 times larger than most other comets. Though comets are usually small, they look much bigger to us because their tails of melting ice reflect light. Only about one comet per year is visible to the human eye.

Meteors are a much more common sight in the night sky. There are several meteors falling to Earth every hour, but you must be watching closely to see one. Meteor showers, when many meteors fall at the same time, come from a comet's orbit around the Sun. When Earth's orbit crosses the path of a comet's orbit, the tail of dust and rocks enters the atmosphere as meteors. This happens several times a year.

Sometimes space objects can cause a lot of damage when they hit Earth. Large meteoroids—too big to burn up as meteors—hit Earth around every 2,000 years, causing a crater like this one.

The Oort Cloud is like a shell or bubble surrounding our solar system. Scientists think it is made up of billions or trillions of icy objects.

German mathematician Johannes Kepler discovered that space objects speed up when they are closer to the Sun. That's because the Sun's gravitational pull is stronger when the object is closer, meaning that the object must move faster to stay in orbit.

How fast are comets?

Most of a comet's orbit is in the outer solar system. As they get closer to the Sun and start to heat up, comets usually speed up. When Comet Halley is farthest away from the Sun it travels at about two-thirds of a mile per second (1 km per second). But when it is near Venus, it speeds up to about 34 miles per second (55 km per second)!

Some comets have short trips around the Sun. Others have very long trips. Comet Halley can be seen from Earth every 75 or 76 years. Comet Swift-Tuttle passes by Earth every 133 years. It was first identified in 1862 and last seen in 1992. In the 1990s, two amateur astronomers identified a new comet, now named for them: Comet Hale-Bopp. Scientists estimate Hale-Bopp will not return for at least another 2,000 years.

DID YOU KNOW?

There is a calendar of meteor showers! A meteor shower repeats each year as Earth's orbit crosses a comet's orbit at around the same point. Meteor showers are often named for the stars or constellations that are close to where the meteors appear in the sky. The annual Perseids Meteor Shower in August is caused by the space dust left behind by Comet Swift-Tuttle.

We use satellites for many purposes, such as communication, collecting data, taking pictures, tracking weather patterns, and operating the ***global positioning system*** *(GPS).*

How do satellites stay in orbit?

Scientists use the word satellite to describe anything that orbits a planet. This means that a moon is a satellite. But satellites are also machines or human-made objects that orbit Earth or another planetary body.

Just like moons, satellites stay in orbit thanks to the force of gravity. They are usually sent into orbit on a rocket. Then, once they exit the atmosphere, the pull of Earth's gravity keeps them from continuing into space.

DID YOU KNOW?

That moving star is really a satellite! Have you ever looked into the night sky and seen a star blinking or moving across the sky? That is one of the many satellites orbiting Earth. One new type of communication satellite is a network of smaller satellites that provide internet access to remote areas. These satellites can look like a string of lights moving across the night sky.

The European Space Agency is testing a "space claw" that can grab a piece of junk and pull it into the atmosphere, where both objects will then burn up like a meteor.

Do satellites ever crash into each other?

There has only been one accidental satellite crash, when two communications satellites collided in 2009 and were destroyed. To keep collisions like this from happening, space agencies and companies keep track of all the satellites. Each new satellite is launched into a unique orbit so that nothing will be in its path.

However, there are now more than 7,500 objects in orbit. Objects in a very high orbit can stay there for thousands of years. But orbits can change—especially those closer to Earth. And satellites eventually break down. The result is space junk, which could collide with other satellites and spacecraft.

The James Webb Space Telescope sends us images of the universe we've never seen before.

CHAPTER 5: QUESTIONS ABOUT SPACE EXPLORATION

How much of space have we explored?

In the 1960s, the United States and the **Soviet Union** competed in a "space race," in which each nation competed to be the first to explore space. This race led to the famous Apollo missions, which sent humans to the Moon. That's as far as humans have traveled in space. But since then, space agencies all over the world have explored space.

One way to explore space is by **uncrewed spacecraft** that travel beyond the Moon and send information back to Earth. The National Aeronautics and Space Administation (NASA) in the U.S. has sent several robotic vehicles called rovers to study Mars. Japan has also sent spacecraft to collect samples from small asteroids and return them to Earth. Another way we explore space is by using very powerful telescopes. The Hubble Space Telescope was launched into space in 1990. In 2021, the James Webb Space Telescope was launched. It now orbits the Sun.

DID YOU KNOW?

Voyager 1 and 2 have traveled to the edge of the solar system. In 1977, NASA launched the Voyager 1 and 2 space probes. Since then, the probes have collected data about the entire solar system. The gravity of one planet helped the probes keep moving to the next. NASA expects to receive information from the probes until 2036. Then, they will be too far away to communicate with Earth.

The James Webb Space Telescope captured this image of gas "cliffs" within the Carina Nebula, about 7,500 light-years from Earth.

What do astronauts do on the International Space Station?

The International Space Station (ISS) has been orbiting Earth for more than 25 years. Astronauts from 18 different countries have visited the station. And the station keeps growing. New pieces are launched into space and assembled by the crew. Different countries build different parts of the station and are responsible for maintaining those parts.

Astronauts on the ISS conduct science experiments. Scientific research from the space station has helped improve medical equipment and treatments, deal with **climate change**, and discover tools and materials that improve our everyday lives.

***Air filtration** technology developed on the ISS was adapted for use on Earth to purify air during the COVID-19 pandemic.*

Can humans live on the Moon?

So far, the only people to walk on the Moon were 12 American astronauts, between 1969 and 1972. They wore space suits with helmets because the Moon has no breathable air. This means that if humans did live on the Moon one day, it would be in sealed habitats.

NASA and its partners have started a new series of Moon missions called Artemis. The goal of Artemis is to build a base on the Moon, where astronauts will conduct science experiments and learn more about how to live in space. Artemis 1 launched in 2022 to test the new Orion spacecraft without a crew. Artemis 1 traveled to the Moon and back safely in 25 days. In the next planned mission, a crew of astronauts will travel in the Orion spacecraft thousands of miles beyond the Moon—the farthest into space any human has traveled.

The Artemis missions will help scientists plan the first crewed mission to Mars.

Can we travel to Mars?

Scientists think it's possible to travel to Mars—but it's a dangerous trip and we aren't quite ready. Uncrewed missions to Mars take about seven months to get there. But spacecraft with people must carry water, food, oxygen, equipment, and fuel for the return trip. This much bigger and heavier spacecraft would probably take nine months just to reach Mars.

Because their orbits are different, the distance between Mars and Earth changes. So the launch date must take advantage of the shortest distance between the planets. The astronauts would also need to wait a few months on Mars for the orbits to align again for the return trip. Adding all this time, NASA estimates a crewed mission to Mars would take two to three years.

Rovers launched by the U.S. and China, pictured here, have explored the surface of Mars.

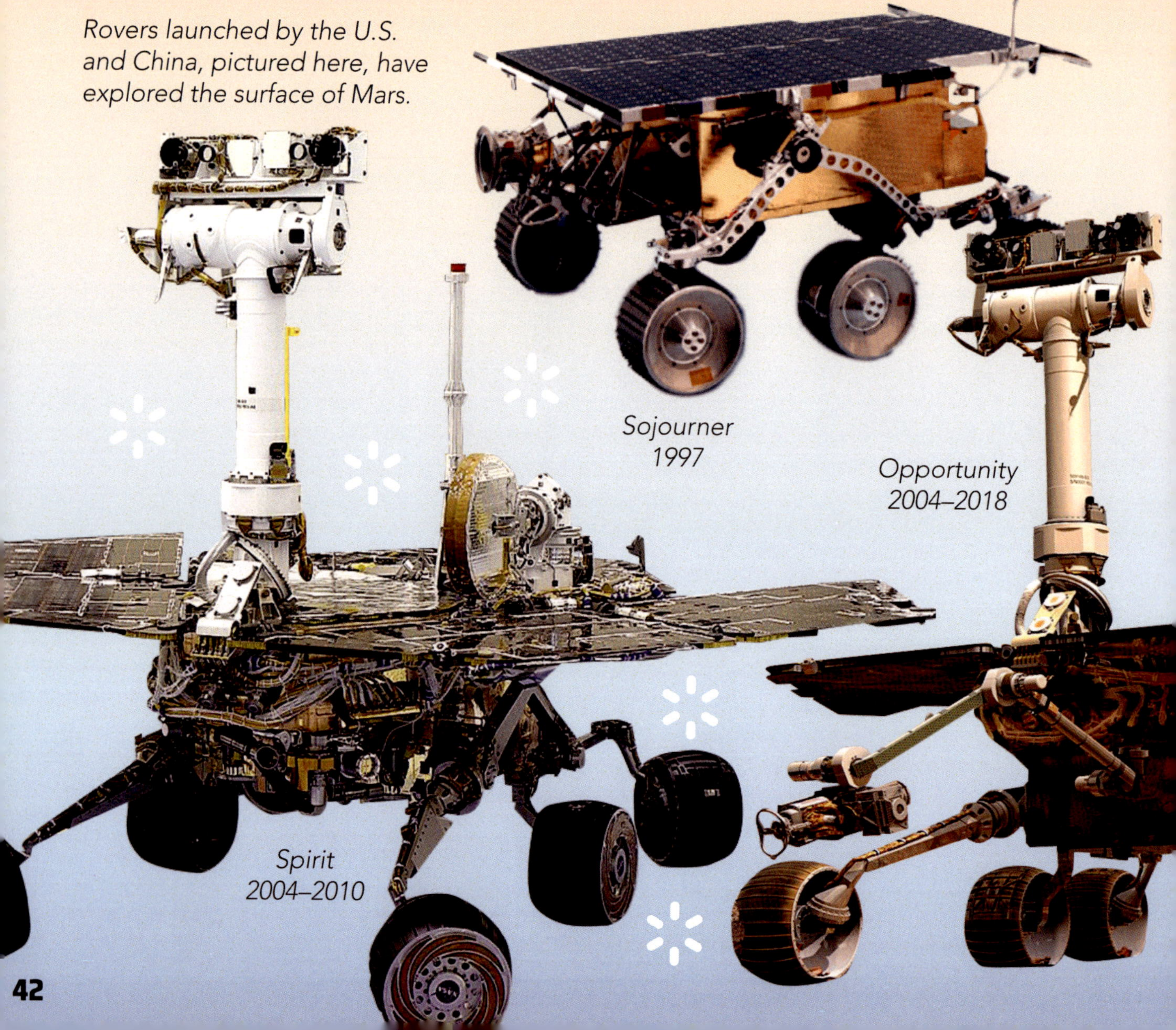

Sojourner 1997

Opportunity 2004–2018

Spirit 2004–2010

The average distance between Earth and Mars is 140 million miles (225 million km).

Curiosity 2012–present

Perseverance 2021–present

Zhurong 2021–2022

Is there life on other planets?

There is no life on other planets in our solar system. But scientists are looking for evidence of life on other planetary bodies with water. Saturn's moon Titan is larger than the planet Mercury, has a dense atmosphere, and has liquid oceans underneath its icy crust. That water may contain **microscopic** life!

So far there is no evidence of life on Titan, but scientists are continuing to study the moon because of its unique environment.

DID YOU KNOW?

There may have been life on Mars billions of years ago! While there is no life on Mars today, the planet may have had an atmosphere and liquid water at one time. Scientists have discovered that some bacteria **here on Earth may be able to survive beneath the surface of Mars. When astronauts visit the red planet, they will test the soil and frozen water to look for bacteria and other signs of ancient life.**

What is a Goldilocks planet?

Scientists have identified over 5,000 exoplanets, or planets outside of our solar system. Most of them are like Jupiter or Neptune, but a few are terrestrial, like Earth and Mars. Only a few hundred of them are in their star's habitable zone. This zone is not too close and not too far from the star to support life. Planets in this zone are called Goldilocks planets because, like the "just right" bowl of porridge in *Goldilocks and the Three Bears*, they are "not too hot" and "not too cold." It's possible that life has evolved on one of these planets, but we are too far away to know for sure.

Earth is a Goldilocks planet. Its conditions are just right to support life.

Glossary

air filtration (air fil-TREY-shuhn): The process of removing particles and pollutants from the air

ammonia (uh-MOHN-yuh): A gas made of hydrogen and nitrogen

asteroids (AS-tuh-roids): Rocky space objects that orbit the Sun

astronomer (uh-STRON-uh-mer): A scientist who studies things beyond Earth's atmosphere

atmosphere (AT-muhs-feer): The layers of gases surrounding a planet

axis (AK-sis): An imaginary line through the center of an object, around which it spins

bacteria (bak-TEER-ee-uh): Single-celled organisms

circumpolar (sur-kuhm-POH-ler): Near one of Earth's poles

climate change (KLAHY-mit cheynj): The long-term change in Earth's usual climate. Global warming is the current period of climate change.

comets (KOM-its): Space objects of ice and rock, with tails of dust and gas, that orbit the Sun

debris (duh-BREE): Pieces of waste

dense (dens): Having a lot of matter packed closely together

diameter (dahy-AM-i-ter): A straight line that passes through the center of a round object

galaxies (GAL-uhk-sees): Systems of stars pulled together by their own gravity

global positioning system (GLOH-buhl puh-ZISH-uhning SIS-tuhm): A system of satellites used to determine one's location on Earth

gravity (GRAV-i-tee): The force that pulls objects toward Earth's center, or that pulls two objects with mass toward each other

high tide (hahy tahyd): The time when the sea reaches its highest level

mass (mas): A measure of the amount of matter in an object

matter (MAT-er): A substance that has mass and takes up space

methane (METH-eyn): A gas made of hydrogen and carbon

microscopic (mahy-kruh-SKOP-ik): Only visible under a microscope

molecules (MOL-uh-kyools): Groups of two or more atoms that make up the smallest units of substances

navigate (NAV-i-geyt): To plan and direct a route

Northern Hemisphere (NAWR-thern HEM-i-sfeer): The half of Earth north of the equator

orbit (AWR-bit): The path of one space object around another

plane (pleyn): A flat surface that extends horizontally in both directions

probes (prohbs): Devices used to gather information

satellites (SAT-l-ahyts): Natural or human-made objects that orbit a larger space object

Southern Hemisphere (SUHTH-ern HEM-i-sfeer): The half of Earth south of the equator

Soviet Union (SOH-vee-it YOON-yuhn): An empire (1922–1991) that included Russia and 14 other countries in eastern Europe and northern Asia

telescope (TEL-uh-skohp): Instruments that make objects appear closer

terrestrial (tuh-RES-tree-uhl): Earth-like; having a hard, rocky surface

uncrewed spacecraft (uhn-KROOD SPEYS-kraft): A spacecraft with no humans on board, that is controlled from the ground

Index

Comprehension Questions

1. **The Kármán Line is about ______ above Earth's surface.**
   a) 62 miles (100 km)
   b) 100 miles (161 km)
   c) 85 miles (137 km)

2. **The ______ constellation was used by sailors in the Southern Hemisphere to help them navigate.**
   a) Big Dipper
   b) Orion
   c) Southern Cross

3. **True or False:** Ceres is the largest object in the asteroid belt.

4. **True or False:** The Kuiper Belt orbits the Sun between Mars and Jupiter.

5. **True or False:** A light-year is a measure of time.

6. **______ is about 300 times the size of Earth.**
   a) Saturn
   b) Jupiter
   c) Uranus

7. **______ are balls of rock, dust, and ice.**
   a) Asteroids
   b) Meteors
   c) Comets

8. **True or False:** One year on Mars is 687 Earth days.

1. a, 2. c, 3. True, 4. False, 5. False, 6. b, 7. c, 8. True

About the Author

Rachel Stuckey is a writer, editor, and learning content designer from Toronto. She has spent much of her life as a digital nomad, working remotely as she travels the world. She has written more than 30 books for young readers.

Crabtree Publishing

crabtreebooks.com 800-387-7650

In Canada: We acknowledge the financial support of the Government of Canada through the Canada Book Fund for our publishing activities.

Hardcover 978-1-0398-8095-5
Paperback 978-1-0398-8456-4
Ebook (pdf) 978-1-0398-8215-7
Epub 978-1-0398-8335-2

Printed in the U.S.A./CP2025

Published in Canada
Crabtree Publishing
616 Welland Avenue
St. Catharines, Ontario
L2M 5V6

Published in the United States
Crabtree Publishing
347 Fifth Avenue
Suite 1402-145
New York, NY 10016

Written by: Rachel Stuckey
Series development: Janine Deschenes
Editorial coordinator: Kathy Middleton
Editor: Janine Deschenes
Proofreader: Melissa Boyce
Art direction and layout: Katherine Kantor and Margaret Salter
Photo research: Katherine Kantor
Production manager: Candice Campbell

Library and Archives Canada Cataloguing in Publication
Available at the Library and Archives Canada

Library of Congress Cataloging-in-Publication Data
Available at the Library of Congress

Images:
European Space Agency: ClearSpace: p. 37 (background)
NASA: p. 11 (top), 33 (inset), 42–43 (Opportunity)
Caltech: p. 39 (top left), 27 (bottom left), 39 (top right)
Caltech SSC: p. 27 (bottom right)
Caltech ASU: 31 (bottom left)
Cory Huston: p. 41
ESA CSA STScl: p. 39 (bottom)
JPL: p. 18 (Neptune), 44 (background)
JPL Caltech: p. 12 (Venus), 15 (Venus), 18 (Uranus), 22 (bottom left)
JPL Caltech: SwRI: MSSS: p. 18–19 (Jupiter)
JPL ESA STScl: p. 27 (top right)
JPL Johns Hopkins University Applied Physics Laboratory:Southwest Research Institute: p. 23
JPL: MSSS: p. 14 (Mars), 43 (top/Mars)
JPL: Space Science Institute: p. 18–19 (Saturn)
JPL: USGS: p. 14 (Mercury)
SETI: P. Jenniskens: p. 31 (bottom right)
The Planetary Society: p. 20–21 (background)
Wikimedia: p. 40 (bottom right), 42 (Sojourner, Spirit), 43 (Caltech, Curiosity, Zhurong)
All other images: Shutterstock